Justin R. Nayagam

Compendium On Costus Pictus: A Medicinal Spiral Ginger

Justin R. Nayagam

Compendium On Costus Pictus: A Medicinal Spiral Ginger

LAP LAMBERT Academic Publishing

Impressum / Imprint
Bibliografische Information der Deutschen Nationalbibliothek: Die Deutsche Nationalbibliothek verzeichnet diese Publikation in der Deutschen Nationalbibliografie; detaillierte bibliografische Daten sind im Internet über http://dnb.d-nb.de abrufbar.
Alle in diesem Buch genannten Marken und Produktnamen unterliegen warenzeichen-, marken- oder patentrechtlichem Schutz bzw. sind Warenzeichen oder eingetragene Warenzeichen der jeweiligen Inhaber. Die Wiedergabe von Marken, Produktnamen, Gebrauchsnamen, Handelsnamen, Warenbezeichnungen u.s.w. in diesem Werk berechtigt auch ohne besondere Kennzeichnung nicht zu der Annahme, dass solche Namen im Sinne der Warenzeichen- und Markenschutzgesetzgebung als frei zu betrachten wären und daher von jedermann benutzt werden dürften.

Bibliographic information published by the Deutsche Nationalbibliothek: The Deutsche Nationalbibliothek lists this publication in the Deutsche Nationalbibliografie; detailed bibliographic data are available in the Internet at http://dnb.d-nb.de.
Any brand names and product names mentioned in this book are subject to trademark, brand or patent protection and are trademarks or registered trademarks of their respective holders. The use of brand names, product names, common names, trade names, product descriptions etc. even without a particular marking in this work is in no way to be construed to mean that such names may be regarded as unrestricted in respect of trademark and brand protection legislation and could thus be used by anyone.

Coverbild / Cover image: www.ingimage.com

Verlag / Publisher:
LAP LAMBERT Academic Publishing
ist ein Imprint der / is a trademark of
OmniScriptum GmbH & Co. KG
Heinrich-Böcking-Str. 6-8, 66121 Saarbrücken, Deutschland / Germany
Email: info@lap-publishing.com

Herstellung: siehe letzte Seite /
Printed at: see last page
ISBN: 978-3-659-62481-0

CONTENTS

LIST OF TABLES

LIST OF FIGURES

CHAPTER 1

COSTUS PICTUS: A PROMISING MEDICINAL SPIRAL GINGER IN HERBAL MEDICINE

1.1. Introduction

1.1.1.Plant Introduction History and its Importance to Mankind

Plant introduction has a global effect, as it is a fundamental occupation of human kind and its discovery dates back to the origin of civilization. It is obvious that earlier activities of hunting and gathering lead mankind to select and propagate plants and domesticate animals by stable communities (Harlan, 1992; Solbrig and Solbrig, 1994). The archaeological investigations as well as historical records of early civilizations in Egypt and the Middle East reveals that the agricultural sector was well organized to produce cereals, vegetables, fruits, dates, flax, cotton and other nonagricultural components of the society (Janick *et. al.,* 1969). Early Chinese writings indicate the knowledge of grafting, layering and other techniques although rice and millet were the principal food sources. It is regarded that Romans developed ornamental gardening to a high level during the period between 500 B.C to A.D. 1000 (Harlan, 1992).

It was during the medieval period (A.D.750 to 1500) the society was organized around large estates and monasteries that acted as independent agricultural and industrial organizations and have preserved a great deal of written and unwritten knowledge. In both kinds of institutions, a separation developed among those involved in the production of cereals, fibres, and forages grown extensively in large fields (agronomy); vegetables, fruits, herbs, and flowers grown in orchards near the home (horticulture); and woody plants (Harlan ,1992) grown for lumbar, fuel and game preserves (forestry). The end of medieval period and the beginning of Modern Europe brought a shift from a subsistence existence to a market economy and the emergence of land ownership in the sixteenth century (Solbrig and Solbrig, 1994).

1

The plant exchanges from the area of origin to developing countries of the world greatly increased the range of plants available for food, medicine, industrial uses, and gardening as well as propagation practices to reproduce them whenever required. In addition to edible food crops, new and exotic plants were sought for introduction that was a result of series of plant exploration trips throughout the world. Captain Cook in 1768 along with Sir Joseph Banks, Francis Masson brought large number of exotic plants to Royal Botanic Garden established at Kew (Hartmann *et. al.*, 1988; Reed, 1942). The Wardian case invented by Nathalial Ward (Ward, 1842) was used in transporting plants through long ocean voyages.

The techniques in gardening, propagation and nursery practices of the time are available in several publications (Baltet, 1910; Fuller, 1887; Bailey, 1920). The concept of plant nursery originated as a part of agriculture. Nevertheless, the development of commercial nurseries is probably something that has developed largely within the recent era (Davidson *et. al.*, 2000). The present day plant propagation industry is large and complex and involves the group that multiplies plants for sale or distribution and also the related industries that provides services, consultations, research and teaching.

1.1.2. Importance of *Costus pictus* as a medicinal plant

The plant *Costus pictus* can be suitably called as a medicinal spiral ginger or insulin plant, is a more recently introduced species to Peninsular India and it is believed to be introduced by nuns in Christian monasteries that acted as independent agricultural organizations and have preserved a great deal of written and unwritten knowledge. Apart from vegetables, fruits, herbs, and flowers grown in orchards they have given interest in growing plants with medicinal interest. The Introduction of *Costus pictus* is one such example. The plant gains more medicinal interest due to the knowledge that their leaves have some medicinal property in curing diabetes in humans. Various

2

research workers have conducted laborious research work to find the active principle in it and many studies are in fast progress in both animal and human models. In this contest, knowing the vegetative and reproductive features gains high importance for identifying the plant material in fresh or preserved conditions when it becomes demanding in the industry as a raw material. In the present investigation due importance have been given to elucidate the morphological and anatomical characters of all the plant parts so that any disputes in identifying the material can be resolved in future by making reference to the present findings.

1.2. Review of Literature

The earliest and may be the first of published records of the taxa *Costus* is found in *Hortus malabaricus* by Hendrick Andariaan van Rheede (1678 – 1693) with respect to Kerala conditions. The family Zingiberaceae consists of around 53 genera and more than 1200 species, distributed mainly in tropics and subtropics with the centre of distribution in the Indo – Malayan region, but extending through tropical Africa to Central and South America (Kress *et al.*, 2002). The family Costaceae includes 4 genera with about 100 species (Larsen *et al.*, 1998). Costacean members are spread throughout the tropics with their centers of diversity in South America and Africa. Zingiberaceae of a greater portion of Kerala, and portions of South India, is found in Rama Rao's flowering plants of Travancore (1914). Among the 33 species listed in the flora, *Costus speciosus* (Koenig) Smith was also included. Many of the earlier floras (Fischer, 1928; Gamble, 1916 – 1935, Sasidharan, 2004) treated the genera *Costus* under family Zingiberaceae. Recently, Costaceae were classified as Costoideae, a subfamily or as tribe (Costeae) within Zingiberaceae. The genus *Costus* along with genera *Dimerocostus*, *Monocostus* and *Tapeinochilus* form the family Zingiberaceae were transferred to the family Costaceae (Nakai, 1941). Zingiberaceae can be easily differentiated from Costaceae as the zingiberacean members have essential oils and distichous arrangement of leaves (Sabu, 2006).

1.3. Methodology

1.3.1. Location of the Study and its Physiography:

Kerala State lies along the south – west corner of the Peninsular India, between 8° 18' and 12° 48' N latitude and 77° 22'E longitude. The boundaries of the State include the Lakshwadeep Sea in the west, Tamilnadu in the south and east and Karnataka in the north. The State has an area of 38,863 km², which is about 1.18 percent of the total area of India and is administratively divided into 14 districts. For the present investigation samples were collected from the existing cultivation plots at Kaintikkara, Kakkanad and Kidangoor, which comes under Ernakulam district of Kerala State, India.

The physiography of the state is highly diversified due to the long tract of Western Ghats on the eastern side of Arabian Sea. The complex topography of the State includes mountains, valleys, ridges and scarps. The altitude varies from sea level to a highest of 2695 m above MSL. Based on altitude increase the land area is classified into high ranges (above 750 m ASL); highlands (between 75 – 750 m ASL); midlands (between 7.5 -75 m ASL) and lowlands (below 7.5m ASL). About 43 percent of the total land area of the State is highland, which is followed by midland, which covers 42 percent, high ranges 15 percent and lowland 10 percent (Kerala Land Use Board, 1997).

1.3.2. Climate of the Collection area

The location of experimental study experiences heavy rainfall during southwest monsoon season followed by northeast monsoon season. During the other months the rainfall is considerably less. March, April and May months are the hottest. December to February months is the coldest.

4

The annual rainfall ranges from 3233 to 3456mm at different places of Ernakulam district. The district received an average 3359.2mm (based on 1901-99 data) of rainfall annually. Rainfall during Southwest monsoon season contributes nearly 67.4% of total rainfall of the year, followed by the northeast monsoon, which contributes nearly 16.6%, and the balance of 16% is received during the month of January to May as summer/pre-monsoon showers.

The mean monthly maximum temperatures range from 28.1 to $31.4^{o}C$ and the minimum ranges from 23.2 to $26^{o}C$. The maximum temperature occurs during March and April months and the minimum temperature occurs during December and January months. The humidity ranges from 68 to 89% during morning hours And 64 to 87% during evening hours. The maximum humidity is observed during May to October months.

The wind speed ranges from 6.7 to 10.9 km/hr with mean speed of 9.1 km/hr. The wind speed is high during the period from March to September. The PET (potential evapo-transpiration) ranges from 94.5 to 159.2mm. The maximum PET occurs during March and minimum occurs during June.

On the basis of morphological features and physico-chemical properties, the soils of Ernakulam district are classified as Lateritic, Hydromorphic saline, Brown hydromorphic, Reverine alluvium and Coastal alluvium. The soil of the study area is Lateritic that are well-drained, low in organic matter and plant nutrients. The major crops grown are coconut, tapioca, rubber, areca nut, pepper, cashew and spices (Shyam, 2007).

1.3.3. Basic information on the study species

Up-to-date nomenclature of all the nine species was worked out in accordance with the International Code of Botanical Nomenclature and a few synonyms are included in the nomenclature part, by which the species are known in National, regional, State and locally. Local names used for the species were gathered from literature and also during field and herbarium studies.

A brief description of each species was prepared based on fresh collections and by literature scrutiny, giving details of habitat, leaves, inflorescence, flowers, fruits and seeds. Data on flowering and fruit ripening periods were also gathered based on field and herbarium data. The distribution pattern of the species in Kerala was gathered from field, herbarium studies and literature studies.

1.3.4. Method in herbarium preparation

Samples of specimens collected from each sample plot after tagging were taken to the lab and then pressed after noting the vegetative features. As the specimen is highly fleshy, additional sheets with blotting quality were placed in between twigs.The blotting papers in the plant press were regularly changed at definite intervals. Once the pressed specimens are ready, they were subjected for poisoning using mercuric chloride solution and then mounded on standard herbarium sheets of the size 29cm x 41 cm and stored in a dry place for future reference.

1.3.5. Specimen collection and processing for morphologic studies

The flowering period of the species is very short and is associated with rainy season when their vegetative parts have luxurious growth, which causes difficulty to enter into the plot even for gathering inflorescence and flowers. The association of inflorescence with ants and insects also makes the job tedious. The specimen collection is rather difficult as the flowers are very delicate and fleshy so that they wither and crumble forming a gummy yellow mass soon after collection, making it difficult to study the floral morphology. Spade and diggers used to lift the rhizome together with their adventitious roots and aerial shoot from the field. Soon after gathering field information the rhizomes covered with adventitious roots were washed thoroughly to remove the soil and then the rhizomes together with roots were packed together whereas aerial shoot and inflorescence were detached and packed separately with labeling before transporting it to the laboratory.

1.3.6. Specimen collection and processing for anatomic studies

For the temporary mount preparation of free hand sections fresh materials such as adventitious roots, aerial shoot, underground rhizome and foliar leaves were collected and soon after sizing with a sharp blade, they were fixed in formalin. Fresh materials using adventitious roots, aerial shoot, underground rhizome, foliar leaves anthers and ovary soon after collection were also used during the present study.

The specimens obtained were washed in distilled water and sectioning using razor is performed. The selected thin sections were then passed to a cavity block containing safranin. The excess stain is washed off by washing in water and then mounted in a microscopic glass slide using glycerin and observed under a light microscope.

1.3.7. Taxonomic position

Costaceae Nakai, J. Jap. Bot. 17: 189. 1941; Tomlison, Evol 16: 197. 1962, Anat. Monocot.3: 360. 1969;

Costoideae (Meisn.) K. Schum. In Engler, Bot. Jahrb.27: 265.1899, Pflanzenr. 4(46): 377. 1904; B.L. Burtt, Notes Roy. Bot. Gard. Edinburgh 31: 162. 1972.

Costeae Meisn., Pl. Vasc. Gen. 2: 291. 1842; Hutchinson, Fam. Fl. Pl ed. 2.2: 584. 1959.

As the family is closely related to family Zingiberaceae, in earlier studies the species under the family were classified under family Zingiberaceae or as subfamily of Zingiberaceae. The genus *Costus* along with genera *Dimerocostus*, *Monocostus* and *Tapeinochilus* form the family Zingiberaceae were transferred to the family Costaceae (Nakai, 1941).

7

Their leaves are spirally arranged and non – aromatic; Leaf sheaths closed on one side, opposite to the lamina. Flower solitary, axillary or spicate. Labellum small or large and showy with wavy surface and characteristic stripes (*Costus pictus*).Lateral staminodes absent. Filament appears petaloid or narrow. External epigynous gland is absent.

Distribution: The family Costaceae includes 4 genera with about 100 species (Larsen *et al.*, 1998) which are distributed to the Neotropics, tropical Africa and some in regions of Asia. *Costus* is the only genus that occurs widely in South India of which *Costus speciosus* is much common as road side plant or growing wildly in openings of plantations, or as undergrowth in dense forests (Sabu, 2006).

Costus L.

Costus L. Sp. Pl. 1: 2. 1753, Gen Pl. ed. 5. 2. 1754; Roxb., Asiat. Res. 11 : 349. 1810, Fl. Indica 1: 58. 1820: Benth. & Hook. f., Gen. Pl. 3: 646.1883: Baker in Hook. f., Fl. Brit. India 6: 249. 1892; K. Schum in Engler, Pflanzenr. 4(46): 378. 1904: B.L. Burtt & R.M. Sm., Notes Roy. Bot. Gard. Edinburgh 31: 200.1972; M. Sabu & Mangaly, Proc. 2[nd]Symp. Fam. Zingiberaceae 17.1996.

Rhizome is perennial, branched, yellowish inside. Stem is erect, branched, perennial, covered with reddish tubular sheaths towards the base, upper tubular sheaths with lamina. Leaves spirally arranged on the stem, shortly petiolate; lamina oblong or oblong – lanceolate, acute or acuminate, often cuspidate, glabrous; ligules short, truncate. Inflorescence appears dense globosely or ellipsoid, which arises terminally on leafy stem. Bracts broad, overlapping at the base reddish towards base and greenish towards outside, each subtends a single flower. Bracteoles smaller and laterally flattened with reddish color. Calyx is tubular, with tip unequally 3-lobed, dorsal broader than the laterals, hard and thorn – like. Corolla tube is shorter or longer than the calyx with its lobes overlapping. Labellum appears large usually, ovate, the edges are often crisped, white or brightly colored. Lateral

8

staminodes are absent. Stamen with a broad filament, which usually curves forward and closes the mouth of the corolla tube unturned tip. Epigynous glands absent; but two nectar secreting cavities are present. Ovary is trilocular; ovules many in 2 rows; style long, filiform but stigma with a crescent-shaped depression, margin ciliate. Capsule are 3-angled, the lateral angles smaller and more spreading, loculicidal splits not reaching to the apex. Seeds ovoid or sub globose, angular with white, fleshy aril; all seeds adhere together by their arils on dehiscence. Embryo is usually straight in copious endosperm, operculum present.

Distribution: Species distribution in Asia is very scanty. Baker (1890-1892) reported *Costus speciosus* from India, but Mass (1979) identified several species from India through herbarium studies.

Ecology: Majority of the species are restricted to the tropical areas, at plains and high altitudes in dense forests, roadside cuttings and plantations.

Phenology: Flowering is associated with the onset of monsoon in Kerala regions and fruit setting is very minimum or even absent in many cases. Flowering and fruiting extends from April – December, and in some species it occurs throughout the year (Sabu, 2006).

Features for identification: The members of Costaceae family can be easily identified from Zingiberacean members by the spiral arrangement of leaves, the closed leaf sheaths, the sunken nectar-glands and the absence of aromatic substance in them.

The taxa *Costus pictus* can be easily identified form other members within the family as their leaves are glabrous on both sides; inflorescence green; labellum creamy yellow with their lateral lobes having reddish brown stripes.

CHAPTER 2

MORPHOLOGICAL AND ANATOMICAL FEATURES OF
COSTUS PICTUS

Costus pictus D. Don, Bot. Mag. T. 1594; Horan. Monogr. 37. 1862; K. Schum. In Engler Pflanzenr. 4(46): 396. 1904.

Costus mexicanus Leibm., Bot. Tidsskr 18: 261. t.16. 1892: K. Schum. In Engler, Pflanzenr. 4(46): 397. 1904.

Common name: Insulin plant

2.1. Collection centers

For the present investigation samples were collected from the existing cultivation plots at Kidangoor with geographic position + 10° 11' 34.99" N, + 76° 24' 14.93"E, with 83 feet elevation (further cited as Plot 1 or P1), Cultivation plot Civil Station Kakkanad (Plot 2 or P2), trial plot at Edayar (Plot 3 or P3), + 10° 05' 10.36" N, + 76° 18' 17.30"E, with 28 feet elevation which comes under Ernakulam district of Kerala State, India. Specimens were also collected from the Botanic garden under the university of Kerala, Karyavattom campus, Thiruvananthapuram (Plot 4 or P4), + 8° 33' 55.63" N, + 76° 53' 10.98"E, with 142 feet elevation, which comes under Thiruvananthapuram district of Kerala State, India, cultivation fields of Farming Trust of India Maruthur road, Palakkad (Plot 5 or P5), + 10° 45' 46.27" N, + 76° 41' 50.05"E, with 349 feet elevation, in Palakkad district, Kerala State, India and from the Rock Gardens, in the foot hills of Kodaikkanal (Plot 6 or P6), + 10° 11' 22.91" N, + 77° 40' 06.36"E, with 1056 feet elevation, in Madurai district, Tamil Nadu State, India.

2.2. Morphological features

2.2.1. Aerial Shoots

Leafy shoot under strong sunlight and at hot climates (P5 & P6) recorded to a height of 35cm to 45cm at flowering and under shade attained a height of 1.15m to 2.4m at flowering (P1, P2 P3 & P4). Spiraling nature was minimum towards the base of shoots (Fig. 2.1.a & Fig. 2.1.b).   .

Fig. 2.1.a. *Costus pictus* – Plants in garden

Fig. 2.1.b. *Costus pictus* – Flowers

Older aerial shoots produces branching and bud formation from leaf axils, which further produces adventitious roots and develops into propagules. As much as 18 such lateral buds were observed for maximum from a plant under shade habitat. They can be separated or pegged into the ground for further growth (Fig. 2.2.a & Fig. 2.2.b).

Fig. 2.2.a. Aerial shoot with axillary buds

Fig. 2.2.b. Buds developing into propagules with adventitious roots

2.2.2. Leaf:

Number of foliage leaves varies from 18 to 25 under strong sunlight and at hot climates (Plot No. 1 and 6) and 23 to 36 leaves per shoot were observed for plants grown under shade at flowering (Plot No. 2, 3 and 4). Leaves are arranged spirally on the erect shoot

Leaf lamina glabrous at both surfaces but varies in shape and length within a shoot and between plants at different habitats. Average length of basal leaves was 6.1 to 7.2cm and a width of 3.1 to 3.4cm from plot no P1, P2, P3 and P4 but 4.1 to 5.6cm from P5 and P6. The longest leaf observed in P2, P3 and P4 had 24.2 to 29.8cm length and width of 8.7 to 11.3cm (Fig. 2.3.a and Fig. 2.3.b). But the average length of basal leaves from plot 5and 6 varied from 5.2 to 6.1cm in length and width ranges from 2.8 to 3.1cm. The longest leaf observed in P5 and P6 had 18cm to 23cm length and width of 6.8 to 8.3cm.

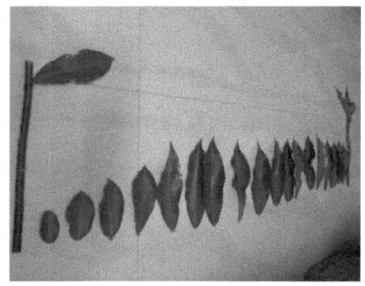 

Fig. 2.3.a. Leaf variations within an aerial shoot   **Fig. 2.3.b. Flowering shoot with smallest and longest leaf**

Basal leaves with are ovate shape with sub cordate base, but leaves towards middle of the shoot have elongated lanceolate lamina with acuminate leaf apex. Leaf margins are entire, with prominent mid ribs and pinnately parallel venation (Fig. 2.4).

Leaves possess ligules with 0.2 to 0.3cm length and a short petiole. The sheathing leaf base covers the erect stem. The Sheaths usually appear reddish in color towards the base of the stem and are greenish towards the tip.

2.2.3. Rhizome and adventitious roots:

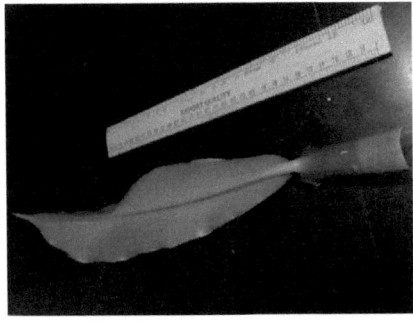

Rhizome measured a longest of 42cm length in a 3 year old plant with 3.5cm stem thickness. As much as 18 aerial shoot and 226 adventitious roots with several lateral branching were observed for maximum from plot 2 and 3 (Fig. 2.5.a – Fig. 2.5.d).

Fig.2.4. Leaf with lamina and leaf base split open

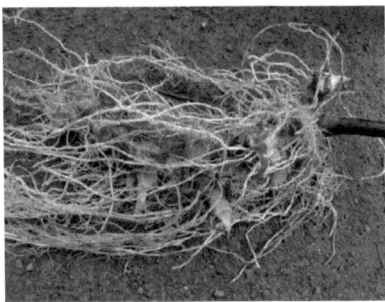

Fig. 2.5.a. Rhizome covered with adventitious roots

Fig. 2.5.b. Rhizome after detaching adventitious roots

Fig. 2.5.c. Rhizomewith nodes and internodes

Fig. 2.5.d. Rhizome with bud

2.2.4. Inflorescence and Flowers:

Inflorescence is terminal in origin, which arises as a sub globose compact spike but sometimes extends and elongates to produce more than 30 flowers from a single spike before it terminates its growth. Peduncle measures a maximum of 5.9cm (P6) to 11.6cm (P1 & 2) with 12 to34 bracts in them. Bracts are numerous which are closely imbricating with a reddish brown coloration in its covered portion but appear greenish in its exposed portion. Bracts at basal portion of the inflorescence appear larger in size with a length and width of 3.4 x 4.1cm and smallest towards apex with 2 .1 x 1.cm (Fig 2.6.a – Fig. 2.6.d).

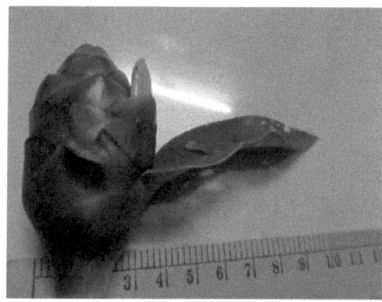

Fig. 2.6.a. An young inflorescence

Fig. 2.6.b. Inflorescence with flowers and detached basal bracts

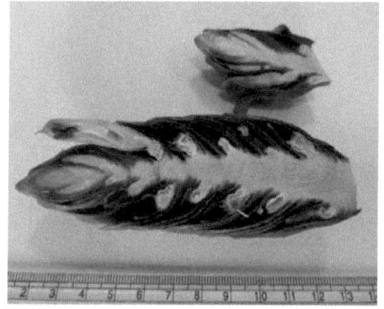

Fig. 2.6.c. L.S. of inflorescence

Fig. 2.6.d. Detached bracts, bracteoles, flower and peduncle of an older inflorescence

Bracteoles appear reddish brown in color which is boat shaped with an average size of 1.6cm length and 1.4cm width. They have more or less pointed tip but with scarious margins. Flower size ranges from 5.1cm (P6) to 6.9cm (P2, P3), which are glabrous and lemon yellow in color with characteristic reddish markings on its labellum. However the flowers from P6 appeared dull yellow with reddish markings.

Calyx with three lobes measures a length of 0.7 – 0.9cm, cup shaped, reddish brown, pubescent and persistent. Corolla tube 0.5 – 0.6cm long, lemon yellow, lobes ranges 4.2 x 1.3 cm to 5.2 x 2.2cm and ovate. Labellum with 3.5 x

4 cm to 5.5 x 4.5 cm, almost formed tubular with three lobes at apex, light yellow and have reddish stripes towards the tip (Fig 2.7.a – Fig. 2.7.f).

Filament of functional stamen appears yellow, broad and hairy towards centre; anther thecae 0.6 x 0.4cm size, creamish white, dehisces by longitudinal slits. Anther crests 1 x 0.5cm, broadly ovate, median yellow with reddish brown margins. Ovary ranges 0.6 x 0.3 x 0.4cm to 0.9 x 0.6 x 0.5cm in dimension. Ovary is trigonous in nature, whitish in color and pubescent. Style terminal in origin and extends to about 4.3cm (Fig 2.8.a & Fig. 2.8.b).

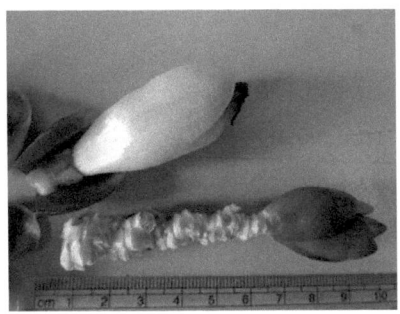

Fig. 2.7.a. Peduncle after detaching flowers and bracts

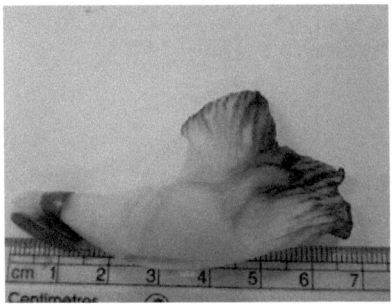

Fig. 2.7.b. Entire flower

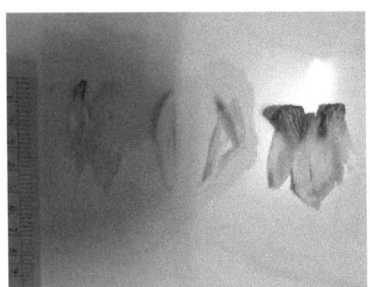

Fig. 2.7.c. Dissected flower showing floral parts

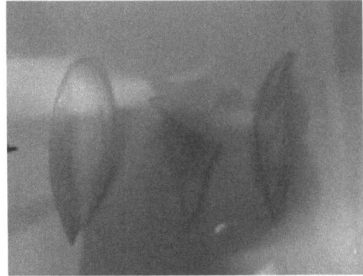

Fig. 2.7.d. Dissected petals

Fig. 2.7.e. L.abellum of flower

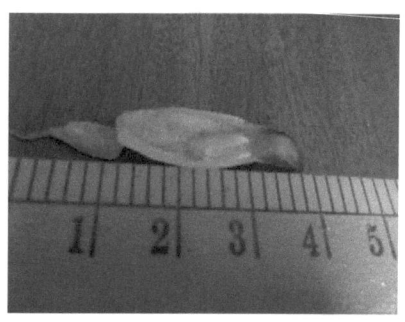

Fig. 2.7.f. Detached functional stamen

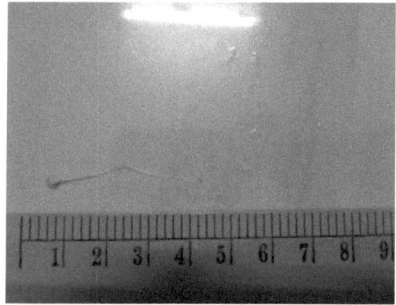

Fig. 2.8.a. Detached stigma with style

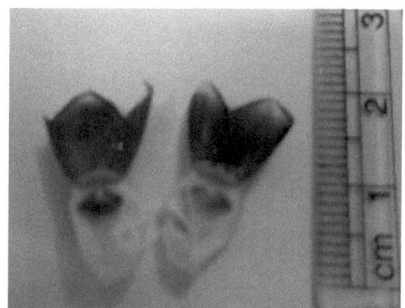

Fig. 2.8.b. Ovary split open

2.3. Herbarium preparation and deposition

The pressed specimens using wooden press were poisoned using mercuric chloride solution and then mounded on standard herbarium sheets of

the size 29cm x 41 cm and after labeling they are stored in dry cupboard for future reference (Fig.2.9). The herbarium samples are also to be deposited to the local, regional, State and National Botanic institutions in India.

Fig. 2.9. Herbarium sheet with specimen
before labeling

2.4. Anatomic features

For the temporary mount preparation of free hand sections fresh materials such as adventitious roots, aerial shoot, underground rhizome and foliar leaves were collected and soon after sizing with a sharp blade, they were fixed in formalin. Fresh materials using adventitious roots, aerial shoot, underground rhizome, foliar leaves anthers and ovary soon after collection were also used during the present study.

The specimens obtained were washed in distilled water and sectioning using razor is performed. The selected thin sections were then passed to a cavity block containing safranin. The excess stain is washed off by washing in water and then mounted in a microscopic glass slide using glycerin and observed under a light microscope.

Anatomic features of the aerial stem, underground rhizome, adventitious roots and foliage leaves were carried out and characteristic features were observed. Aerial shoot and underground rhizome show similarity in its basic structure with the presence of conjoint collateral and closed vascular bundles, which are scattered throughout the ground tissue.

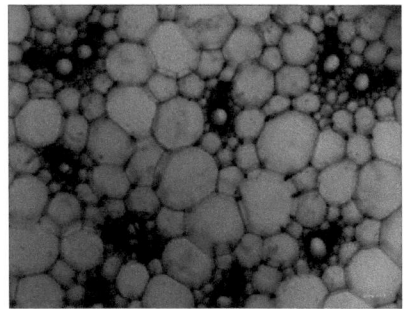
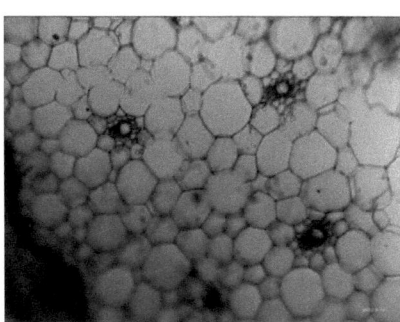

Fig. 2.10.a. C.S. of tender aerial stem   **Fig.2.10.b. T.S. of young aerial stem**

Presence of calcium oxylate crystals is characteristic in the parenchymatous ground tissue, which is smaller in size towards the tip of aerial shoot (Fig. 2.10.a & Fig. 2.10.b) but bigger towards the base of the stem. The crystal size in underground rhizome was found comparatively bigger than those in aerial shoot (Fig. 2.11.a. & Fig.2.11.b)

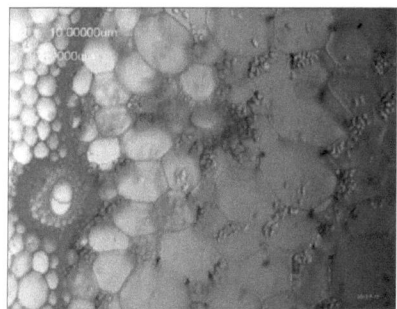

Fig. 2.11.a. T.S. of aerial stem with Oxylate crystals

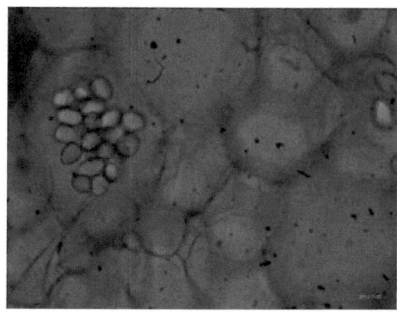

Fig. 2.11.b. Portion of Rhizome with oxylate crystals

The sheathing leaf base on aerial stem almost encircles the shoot, which begins from one node to and sheaths upwards almost to the next node. Sheaths are 8 -10 layered parenchymatous, with vascular traces but with characteristic presence of trichomes (Fig. 2.12.a & Fig. 2.12.b)

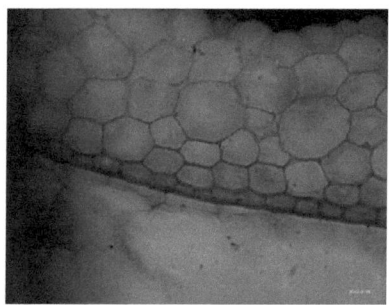

Fig. 2.12.a. T.S. of Sheathing leaf base

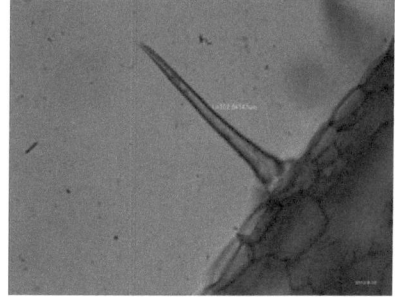

Fig. 2.12.b. T.S. of Sheathing leaf base with trichomes

Adventitious roots possess epidermis and complete parenchymatous cortex with exarch and polyarch xylem. Some cells in the inner cortex possess storage cells, which appear reddish brown when stained with safranin. Vascular bundles are radial with prominent pith made of parenchyma cells (Fig. 2.13.a & Fig. 2.13.b.).

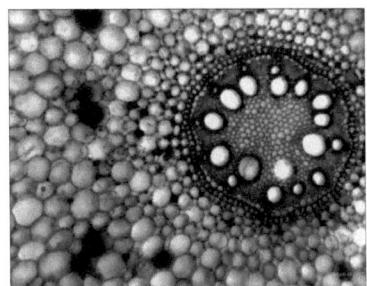

Fig. 2.13.a. T.S. of Adventitious root
– Portion with stele

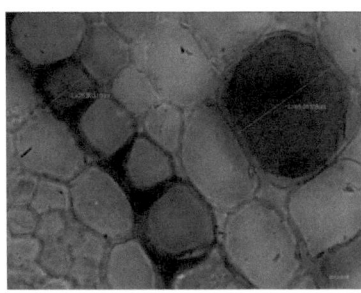

Fig. 2.13.b. Portion of inner
cortex of root with storage cell

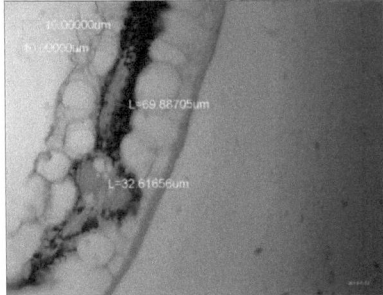

Fig. 2.14.a. T.S. of lamina
– Portion enlarged

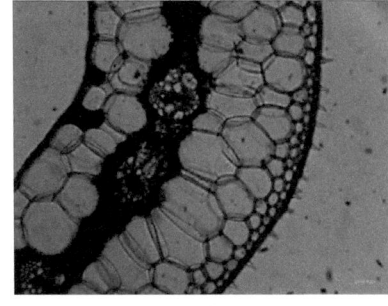

Fig. 2.14.b. T.S. of Leaf – Portion
enlarged through main vein

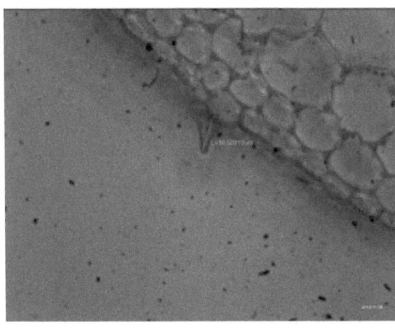

Fig. 2.14.c. Portion of sheathing leaf

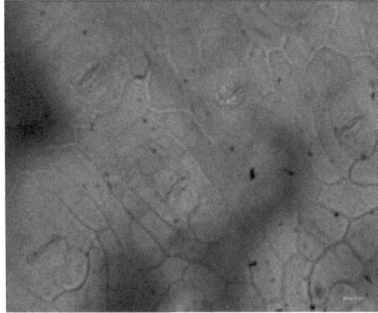

Fig. 2.14.d. Epidermal base with hair
peel with stomata

Foliage leaves possess adaxial and abaxial epidermis with characteristic broad based stiff hairs arising as spikes towards the central portion of abaxial epidermis. The stomata appear to be tetracyclic.

Mesophyll tissue begins with few layers of blank cells and compactly arranged 8 -10 layers of photosynthetic cells which are small in shape.

Vascular traces have endarch structure with conjoint and closed vascular bundles. (Fig. 2.14.a & Fig. 2.14.d).

By observing the results obtained from various portion of plant collected from different locality, a taxonomist can easily come to the conclusion that the plant described through the present investigation is *Costus pictus*. The plant is explained as a recent introduction from America as a herbal cure for diabetes hence commonly called as 'insulin plant' (Sasidharan, 2004, Sabu, 2006). This is often confused with the more native *Costus speciosus* in the vegetative stage but can be easily differentiated from it.

CHAPTER 3

MORPHOLOGICAL AND ANATOMICAL FEATURES OF *COSTUS SPECIOSUS*: A HIGHLY CONFUSED SPECIES IN MATERIAL IDENTIFICATION AND A POTENTIAL ADULTERANT

3.1. Plant collection, identification and taxonomic studies

Samples of *Costus speciosus* were obtained from their natural habitats in selected area which include Edayar, + 10° 05' 10.36" N, + 76° 18' 17.30"E, with 28 feet elevation which comes under Ernakulam district of Kerala State, Mahatma Gandhi University campus, Priyadarsini Hills, Kottayam, Little flower convent, Munambum of Ernakulam District and from Adimaly hill area.

Costus speciosus (*J. Koen*) Smith, Trans. Linn. Soc. London 1: 249.1791; Roxb., Fl. Ind. 1: 58. 1832; Wight, Ic. T. 2014. Baker in 1853; Hook. f. Fl. Brit. India 6:249. 1892; Fischer and Gamble, Fl. Pres. Madras 1490.1928; Gandhi in Sald.&Nicols., Fl. Hassan Dist. 770. 1976; Matthew & Britto in Matthew, Fl. Tam.Carnatic 3: 1613.1983; Burtt & Smith in Dassan.& Fosb., Rev. Handb. Fl. Ceylon 4: 491. 1983; Nair & Nayar, Fl. Courtallum 2: 390. 1987; Nicols., Suresh & Manilal, An Interpr. Hort. Malab. 316. 1988; Mohanan & Henry, Fl. Thiruvananthapuram 475. 1994.

Banksea speciosa J. Koen. In Retz., Obs. Bot. 3: 75. 1783.

Tsjana- kua Rheede, Hort. Malab. 11: 15 – 16, t. 8. 1692.

Common name: Anakuva

3.2. Vegetative and reproductive morphology of *Costus speciosus*

Leafy shoot under strong sunlight and at hot climates recorded to a height of 25 to 40cm at flowering and under shade attained a height of 68 to 95cm at flowering Spiraling nature was much pronounced towards the terminal portion of shoot (Fig. 3.1.a & Fig. 3.1.b). Bud formation from older shoot aerially is absent.

Fig. 3.1.a. *Costus speciosus* – Leafy shoot
With Inflorescence and flower bud

Fig. 3.1.b. *Costus speciosus* –
Inflorescence

3.2.1. Leaf:

Number of foliage leaves varies from 7 to 12 under strong sunlight and at hot and coastal climates and 10 to 18 leaves per shoot were observed for plants grown under shade at flowering. Leaves are arranged spirally on the erect shoot. Lamina deep green, less glabrous on both surface but varies in shape from oblong to lanceolate. Average length of aerial foliage leaves measure 12.5cm length and 3.89cm width. Leaves measured a maximum size of 27 – 10.6cm.

3.2.2. Rhizome and adventitious roots:

Rhizome measured a longest of 33cm length in a 3 year old plant with 2.8cm stem thickness. As much as 9 aerial shoot were observed for maximum.

3.2.3. Inflorescence and Flowers:

Inflorescence is terminal in origin, which arises as a sub globose compact spike, which has a maximum measure of 15cm length and 5.2cm width (Fig. 3.2.b).

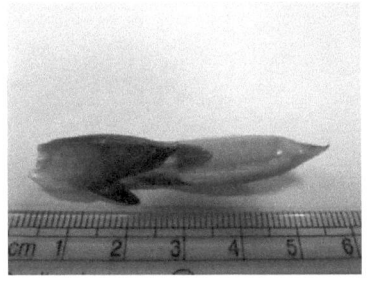 

Fig. 3.2.a. Flower bud with bract & bracteole   Fig. 3.2.b. Entire Flower

Bracteoles appear deep reddish brown in color, less boat shaped with abrupt pointed end and have an average size of 1.9cm length and 0.8cm width. Flower bud flesh colored with pointed ends. Flower size ranges from 5.1cm to 8.1cm, which is less glabrous and creamy white in color (Fig 3.2.a – Fig. 3.2.b).

Calyx with three lobes measures a length of 2.3 – 0.5cm, reddish brown and persistent. Corolla tube 0.5 – 0.6cm long, creamy white, lobes ranges 4.8 x 1.6 cm to 5.5 x 1.9cm and ovate. Labellum has an average size of 7.6 x 6.8cm (Fig 3.3.).

Filament of functional stamen appears light yellow, broad and hairy towards centre. Ovary - tricarpellary, syncarpous, inferior with axile placentation. Style - terminal with an average of 4.6cm length and stigma lobed with 0.25cm wide.

Fig. 3.3. Parts of a dissected Flower

Fruit and seed formation is present in Kerala conditions even though it is believed to be an introduced species under Kerala conditions. Fruit is loculicidal capsule with persistent calyx lobe with an average size of 4.4 x 1.6cm. Seeds black with white aril and have angular ends, almost square shaped or its variants. Seeds measured a maximum of 0.42 x 0.25cm size (Fig. 3.4. a & b).

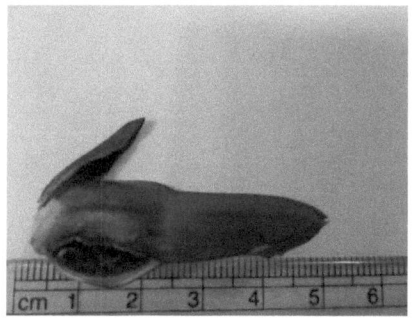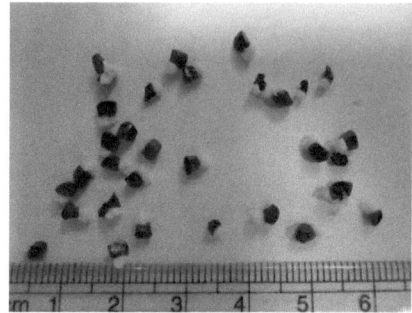

Fig. 3.4. (a & b) Fruits and seeds of *Costus speciosus*

3.3. Anatomic Features of Vegetative Parts

Anatomic features of the aerial stem, underground rhizome, adventitious roots and foliage leaves were carried out and characteristic features were observed. Aerial shoot and underground rhizome show similarity in its basic structure with the presence of conjoint collateral and closed vascular bundles which are scattered throughout the ground tissue.

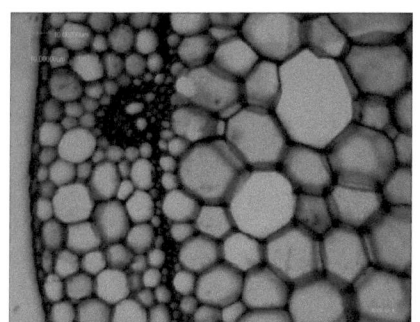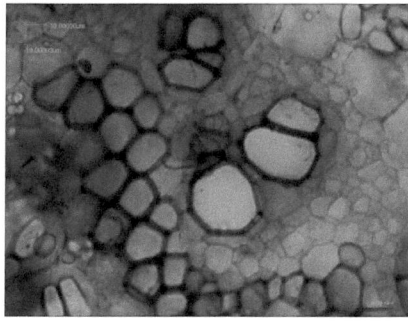

Fig. 3.5. (a & b) T.S. of young aerial stem

Presence of calcium oxylate crystals is characteristic in the parenchymatous ground tissue, which is smaller in size towards the tip of aerial shoot but bigger towards the base of the stem (Fig. 3.5 a & b).

Rhizome of *Costus speciosus* has characteristic outer cortex with a few layers of more or less rectangular cells. The calcium oxalate crystals are bigger in size and may be oval to elongated type (Fig. 3.6.a & Fig. 3.6.b).

24

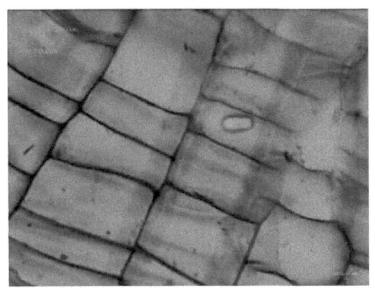
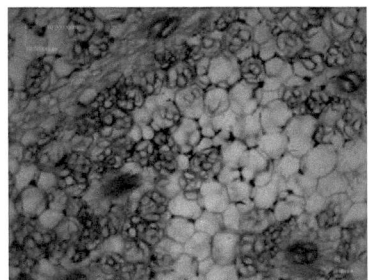

Fig. 3.6.a. T.S. of Rhizome – outer cortex   Fig. 3.6.b. Portion of Rhizome with calciumoxylate crystals

Adventitious roots possess epidermis and complete parenchymatous cortex with exarch and polyarch xylem. Vascular bundles are radial with prominent pith. Root hairs originate from the outer surface of root epidermis and lateral roots arise from pericycle. Metaxylem characteristically shaped with more or less a trigonous outline (Fig. 3.7.a & Fig. 3.7.b).

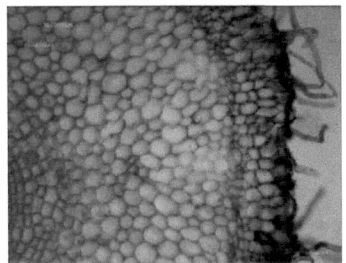
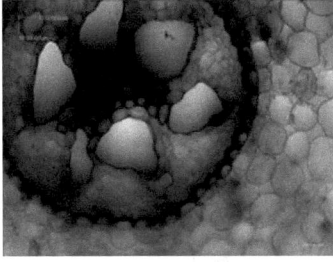

Fig. 3.7.a. T.S. of Adventitious root – Portion   Fig. 3.7.b. Portion with Stele Without stele

Foliage leaves possess adaxial and abaxial epidermis with characteristic broad based stiff hairs arising as spikes. The stomata appear to be tetracyclic. Mesophyll tissue begins with few layers of blank cells and compactly arranged

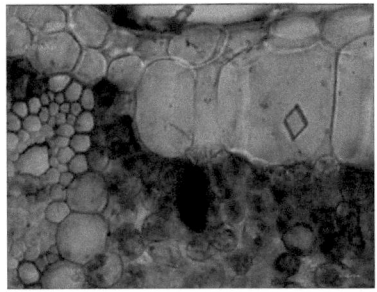

8 -10 layers of photosynthetic cells which are small in shape. The presence of cubical calcium oxalate crystals is a characteristic feature. Vascular traces have endarch structure with conjoint and closed vascular bundles. (Fig. 3.8).

Fig. 3.8. T.S. of Leaf with Characteristically shaped Crystals

Based on the above morphological and anatomical features the spiral ginger *Costus speciosus* can be identified taxonomically from the medicinal spiral ginger *Costus pictus* medicinal using it as raw drug in herbal medicine industry.

CHAPTER 4

COMPARATIVE TAXONOMIC STUDIES OF *COSTUS PICTUS* AND *COSTUS SPECIOSUS*

4.1. Botanical Nomenclature and Common Names of *Costus pictus* and *Costus speciosus*

Costus pictus D. Don, Bot. Mag. T. 1594; Horan. Monogr. 37. 1862; K. Schum. In Engler Pflanzenr. 4(46): 396. 1904.

Costus mexicanus Leibm., Bot. Tidsskr 18: 261. t.16. 1892: K. Schum. In Engler, Pflanzenr. 4(46): 397. 1904.

Common name: Insulin plant

Costus speciosus (J. Koen) Smith, Trans. Linn. Soc. London 1: 249.1791; Roxb., Fl. Ind. 1: 58. 1832; Wight, Ic. T. 2014. Baker in 1853; Hook. f. Fl. Brit. India 6:249. 1892; Fischer and Gamble, Fl. Pres. Madras 1490.1928; Gandhi in Sald. & Nicols., Fl. Hassan Dist. 770. 1976; Matthew & Britto in Matthew, Fl. Tam. Carnatic 3: 1613.1983; Burtt & Smith in Dassan. & Fosb., Rev. Handb. Fl. Ceylon 4: 491. 1983; Nair & Nayar, Fl. Courtallum 2: 390. 1987; Nicols., Suresh & Manilal, An Interpr. Hort. Malab. 316. 1988; Mohanan & Henry, Fl. Thiruvananthapuram 475. 1994.

Banksea speciosa J. Koen. In Retz., Obs. Bot. 3: 75. 1783.

Tsjana- kua Rheede, Hort. Malab. 11: 15 – 16, t. 8. 1692.

Common name: Anakuva

4.2. Comparative vegetative and Reproductive Morphology of *Costus pictus and Costus speciosus*

4.2.1. Features of Aerial Shoots in *Costus pictus and Costus speciosus*

Leafy shoots of *Costus pictus* under strong sunlight and at hot climates recorded to a height of 35cm to 45cm at flowering and under shade attained a height of 1.15m to 2.4m at flowering Spiraling nature was minimum towards the base of shoots and sometimes the feature is absent (Fig. 4.1). The aerial shoot of Co*stus speciosus* under strong sunlight and hot climates recorded 25 to 40cm height at flowering (Fig. 4.2) and under shade attained a height of 68 to 95cm at flowering. Spiraling nature was much pronounced towards the terminal portion of shoot.

Older aerial shoots produces branching and bud formation from leaf axils, which further produces adventitious roots and develops into propagules. As much as 18 such lateral buds were observed for maximum from a plant under shade habitat. They can be separated or pegged into the ground for further growth (Fig. 4.3.a & Fig.4.3.b). But in *Costus speciosus* the feature of lateral bud formation form older shoot aerially is completely absent.

Fig. 4.1 *Costus pictus* – Plants in garden

Fig. 4.2 *Costus speciosus* –
Plants in their Natural habitat

Fig. 4.3.a. Aerial shoot with axillary buds

Fig.4.3.b. Propagule from lateral bud

4.2.1.1.Leaf:

In *Costus pictus*, number of foliage leaves varies from 18 to 25 under strong sunlight and at hot climates and 23 to 36 leaves per shoot were observed for plants grown under shade at flowering Leaves are arranged spirally on the erect shoot.Leaf lamina glabrous at both surfaces but varies in shape and length within a shoot and between plants at different habitats (Fig. 4.4.a). The longest leaf observed recorded an average of 18cm to 23cm length and width of 6.8 to 8.3cm.

In *Costus speciosus* number of foliage leaves varies from 7 to 12 under strong sunlight and at hot and coastal climates and 10 to 18 leaves per shoot were observed for plants grown under shade at flowering. Leaves are arranged spirally on the erect shoot. Lamina deep green, less glabrous on both surface but varies in shape from oblong to lanceolate. Average length of aerial foliage leaves measure 12.5cm length and 3.89cm width. Leaves measured a maximum size of 27cm length and 10.6cm width (Fig. 4.4.b).

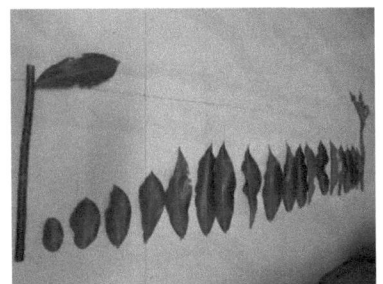 

Fig. 4.4.a. Leaf variations – *Costus pictus*   **Fig. 4.4.b. Leaf variations – *Costus speciosus***

Basal leaves are ovate with sub cordate base, but leaves towards middle of the shoot have elongated lanceolate lamina with acuminate leaf apex, whereas variation in leaf lamina within the shoot is much reduced in *Costus speciosus.* Leaf margins are entire, with prominent mid ribs and pinnately parallel venation are observed in both the species.In both the species sheathing leaf base covers the erect stem and they appear reddish

29

brown to reddish in color towards the base of the stem and are greenish towards the tip.

4.2.1.2. Rhizome and adventitious roots:

In *Costus pictus*, Rhizome measured a longest of 42cm length in a 3 year old plant with 3.5cm stem thickness whereas in *Costus speciosus*, rhizome measured a longest of 33cm length in a 3 year old plant with 2.8cm stem thickness (Fig.4.5.a & Fig.4.5.b). As much as 18 aerial shoot and 226 adventitious roots with several lateral branching were observed for maximum in *Costus pictus*, but in *Costus speciosus* only 9 aerial shoot were observed for maximum.

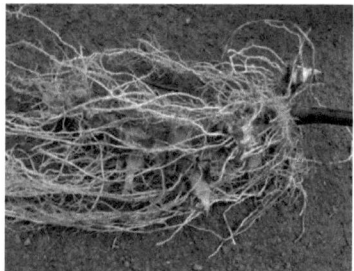

Fig. 4.5.a. Rhizome of *Costus pictus*   Fig. 4.5.b. Rhizome of *Costus speciosus*

4.2.1.3. Inflorescence and Flowers:

Inflorescence is terminal in origin in both the study species, which arises as a sub globose compact spike but sometimes extends and elongates to produce more than 30 flowers from a single spike before it terminates its growth. Bracts are numerous which are closely imbricating with a reddish brown coloration in its covered portion but appear greenish in its exposed portion. Bracts at basal portion of the inflorescence appear larger in size with a length and width of 3.4 x 4.1cm and smallest towards apex with 2 .1 x 1.cm.

Fig. 4.6.a. Inflorescence – *Costus pictus* Fig. 4.6.b. Inflorescence – *Costus speciosus*

In *Costus pictus*, Bracteoles appear reddish brown in color, which is boat shaped with an average size of 1.6cm length and 1.4cm width. They have more or less pointed tip but with scarious margin. In *Costus speciosus*, bracteoles appear deep reddish brown in color which is less boat shaped with abrupt pointed end and have an average size of 1.9cm length and 0.8cm width (Fig. 4.6.a & Fig. 4.6.b).

Flower size in *Costus pictus* ranges from 5.1cm to 6.9cm (Fig. 4.7.a & Fig. 4.7.b), which are glabrous and lemon yellow in color with characteristic reddish markings on its labellum but in *Costus speciosus*, flower buds appear flesh colored with pointed ends and their flower are less glabrous and creamy white in color with size ranges from 5.1cm to 8.1cm (Fig. 4.8.a & Fig. 4.8.b).

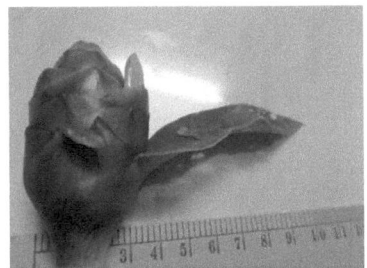 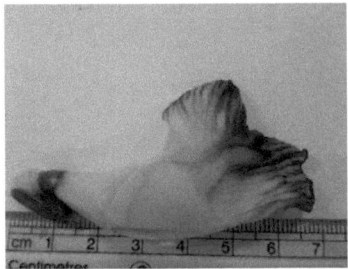

Fig. 4.7.a. Inflorescence with flower bud   Fig. 4.7.b. Entire Flower *Costuspictus*

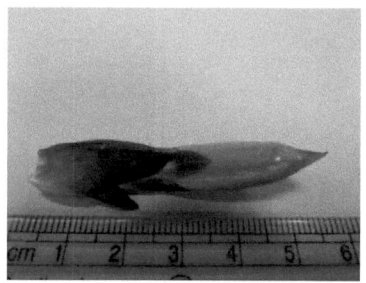 

Fig. 4.8.a. Flower bud   **Fig. 4.8.b. Entire Flower – *Costus speciosus***

In *Costus pictus*, calyx with three lobes measures a length of 0.7 – 0.9cm, cup shaped, reddish brown, pubescent and persistent and that in *Costus speciosus* with three lobes measures a length of 2.3 – 0.5cm, reddish brown and persistent. Corolla tube 0.5 – 0.6cm long, lemon yellow, lobes ranges 4.2 x 1.3 cm to 5.2 x 2.2cm and ovate. Labellum with 3.5 x 4 cm to 5.5 x 4.5 cm, almost formed tubular with three lobes at apex, light yellow and have reddish stripes towards the tip in *Costus pictus* and in *Costus speciosus*, corolla tube is 0.5 – 0.6cm long, creamy white, lobes ranges 4.8 x 1.6 cm to 5.5 x 1.9cm and ovate. Labellum has an average size of 7.6 x 6.8cm (Fig. 4.9.a & Fig. 4.9.b).

In *Costus pictus*, filament of functional stamen appears yellow, broad and hairy towards centre; anther thecae 0.6 x 0.4cm size, creamish white, dehisces by longitudinal slits. Anther crests 1 x 0.5cm, broadly ovate, median yellow with reddish brown margins. Ovary ranges 0.6 x 0.3 x 0.4cm to 0.9 x 0.6 x 0.5cm in dimension. Ovary is trigonous in nature, whitish in color and pubescent. Style terminal in origin and extends to about 4.3cm. In *Costus speciosus,* filaments of functional stamen appear light yellow, broad and hairy towards centre. Ovary tricarpellary, syncarpous, inferior with axile placentation. Style terminal with an average of 4.6cm length and stigma lobed with 0.25cm wide (Fig. 4.10.a & Fig. 4.10.b).

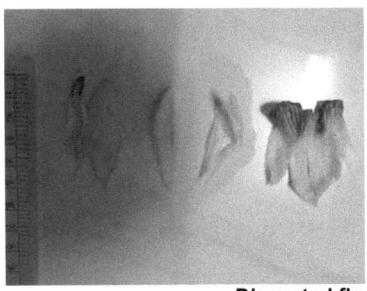 

Dissected flowers showing floral parts

Fig. 4.9.a. *Costus pictus*   Fig. 4.9.b. *Costus speciosus*

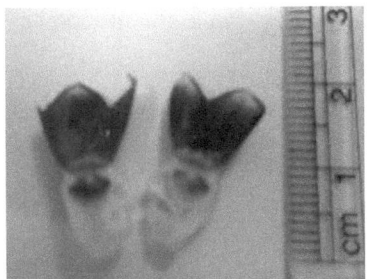 

Ovary split open

Fig. 4.10.a. *Costus pictus*   Fig. 4.10.b. *Costus speciosus*

In *Costus pictus* fruit and seed formation is not reported so far under Kerala conditions but well defined fruit and seed production is observed in *Costus speciosus*. Fruits are loculicidal capsule with persistent calyx lobe with an average size of 4.4 x 1.6cm.

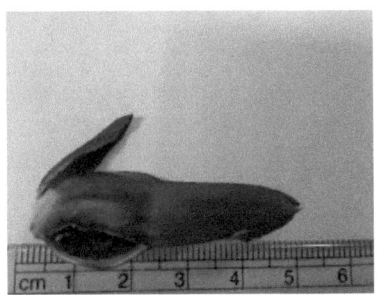 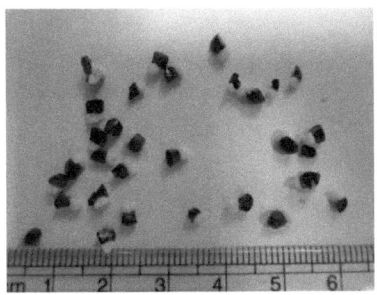

Fig. 4.11 Fruit and Seeds of *Costus speciosus*

Seeds black with white aril and have angular ends, almost square shaped or its variants. Seeds measured a maximum of 0.42 x 0.25cm size (Fig. 4.11).

4.3. Comparative Anatomy of *Costus pictus and Costus speciosus*

Anatomic features of the aerial stem, underground rhizome, adventitious roots and foliage leaves were carried out and characteristic features were observed in both the study species. Aerial shoot and underground rhizome show similarity in its basic structure with the presence of conjoint collateral and closed vascular bundles, which are scattered throughout the ground tissue.

Presence of calcium oxalate crystals is characteristic in the ground tissue, which is smaller in size towards the tip of aerial shoot but bigger towards the base of the stem. The crystal size in underground rhizome was found comparatively bigger than those in aerial shoot in both the study species (Fig. 4.12.a & Fig. 4.12.b). Rhizome of *Costus speciosus* has characteristic outer cortex with a few layers of more or less rectangular cells (Fig. 4.13.a & Fig. 4.13.b).

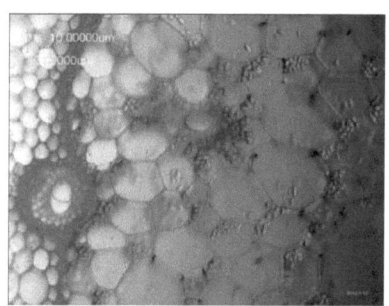 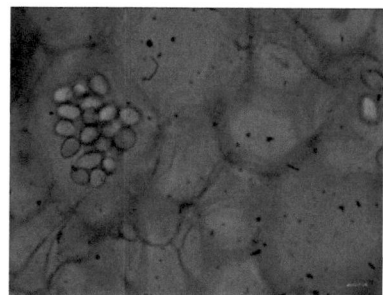

Costus pictus

Fig. 4.12.a. T.S. of aerial stem with crystals   Fig. 4.12.b. Portion of Rhizome with crystals

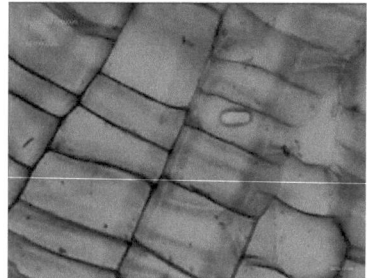 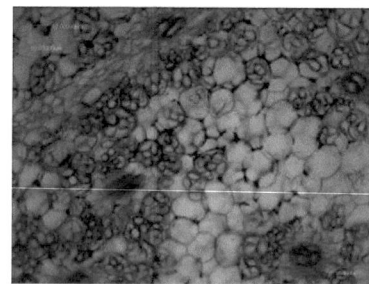

Costus speciosus

Fig. 4.13.a. T.S. of Rhizome – outer cortex   Fig. 4.13.b. Portion of Rhizome with   crystals

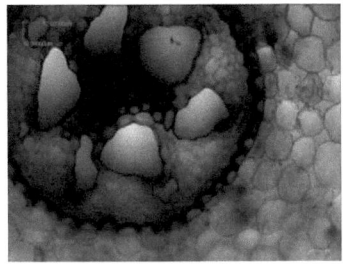

Costus pictus

Costus speciosus

Fig. 4.14.T.S. of Adventitious root with stele   **Fig. 4.15. Portion with Stele**

Roots are adventitious in both the study species and possess an outer epidermis followed by completely parenchymatous cortex with exarch and polyarch xylem in their stele, vascular bundles are radial with prominent pith made of parenchyma cells (Fig. 4.14.). In *Costus speciosus,* metaxylem has characteristic shape with more or less a trigonous outline (Fig. 4.15.).

Foliage leaves in *Costus pictus* possess adaxial and abaxial epidermis with characteristic broad based stiff hairs arising as spikes towards the central portion of abaxial epidermis. The stomata appear to be tetracyclic (Fig. 4.16.a & Fig. 4.16.b).

In *Costus speciosus* foliage leaves possess adaxial and abaxial epidermis with characteristic broad based stiff hairs arising as spikes. The stomata appear to be tetracyclic. Mesophyll tissue begins with few layers of blank cells and compactly arranged 8 -10 layers of photosynthetic cells which are small in shape.

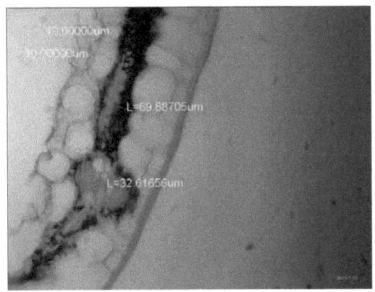
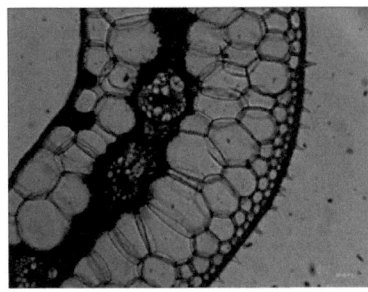

Costus pictus

Fig. 4.16.a. T.S. of lamina – Portion enlarged   **Fig. 4.16.b. T.S. of Leaf – Portion enlarged**

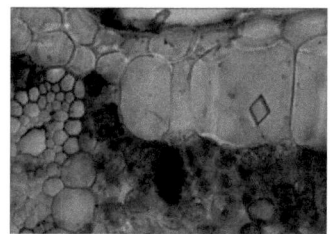

The presence of cubical calcium oxalate crystals is a characteristic feature. Vascular traces have endarch structure with conjoint and closed vascular bundles (Fig. 4.17).

Fig. 4.17. T.S. of *Costus speciosus* Leaf with Characteristically shaped Crystals

4.4. Comparative Herbarium studies of *Costus pictus and Costus speciosus* and its Authentication

The herbarium specimen prepared during the taxonomic studies of *Costus pictus* with collection numbers CP-101 and JN-101 were further processed by poisoning, mounting and labeling and then deposited in the National Herbarium with accession numbers 173772 and 173773 (Madras Herbarium) maintained by Botanical Survey of India (BSI) – Southern Regional Circle, Coimbatore (Fig. 4.18). The *Costus speciosus* specimen maintained in the Botanical Survey of India – Southern Regional Circle, Coimbatore were critically studied (Fig. 4.19) and the herbarium of the collected specimen from the field was prepared into herbarium specimen with collection number CS – 101 and is now maintained in the study centre.

Fig. 4.18. Herbarium specimen of *C. pictus*   Fig. 4.19.Herbarium specimen of *C. speciosus*

The vegetative leaves in herbaria appear coarser in *Costus speciosus* than that in *Costus pictus*. The more distinguishable upper and lower surfaces with depth in color of leaves is characteristic in *Costus speciosus*. The leaves appear much smooth in *Costus pictus* with more or less similar upper and lower halves.

CHAPTER 5

CULTIVATION TRIALS IN *COSTUS PICTUS* USING DIFFERENT ROOTING MEDIUM AND VEGETATIVE PARTS

In order to develop cultivation practice for *Costus pictus* a green house was constructed with a measure of 12.5 x 12.5m^2 and raised standard nursery beds were prepared in which trials using different rooting media was conducted (Fig. 5.1.a & Fig. 5.1.b). The nursery is provided with misting system in order to ensure uninterrupted water supply and to maintain an average temperature ranges 25°C – 35°C and an average humidity that ranges 70% - 85% throughout the study period.

Fig. 5.1.a. Green house for nursery trials   **Fig. 5.1.b Standard nursery beds**

For conduction of nursery trials rhizomes of mother plant, aerial shoot cuttings and axillary bud that arises from older aerial shoots were used and planting material. After gathering the planting material they were repeatedly rinsed in running water and after sizing they were directly planted in equal distance in the nursery bed or in polythene bags with various potting media (Fig. 5.2.a, Fig. 5.2.b, Fig. 5.2.c & Fig. 5.2.d).

Fig. 5.2.a Polythene bags with potting media Fig. 5.2.b Sized aerial shoot cuttings & trichomes

Fig. 5.2.c. Shoot cuttings planted in Nursery beds   Fig. 5.2.d. Propagules for planting

Growth data from different cultivation trials were gathered so that a viable method with sustainable harvest can be suggested.

5.1. Regeneration Characteristics of Planting Materials in Various Planting Medium

Effect of sprouting by the different planting materials in raised standard nursery beds and polythene bags with various sprouting medium was performed. Prefilled polythene bags of the size 22 x 17 cm were utilized for this purpose.

5.1.1. Sprouting Characteristics of Planting Material in Standard Nursery Beds

A total of 69.76 per cent sprouting was obtained in nursery bed with rhizome as planting material. Sprouting was initiated in just 20 days and extended up to 31 days for completion. Maximum percent of sprouting was observed on 23rd day with a record of 24.4% sprouting.

87.55% sprouting was obtained in nursery bed when aerial shoot cuttings were used as planting material. Sprouting was initiated in just 18 days and was completed in 39 days. Maximum percent of sprouting was observed on 24th day with a record of 32.7% sprouting.

When propagules directly obtained from aerial shoot was used as planting material in standard nursery bed 70.24% regeneration was obtained. It took 36 days for the first sprouting to occur and was completed in 52 days.

5.1.2. Sprouting Characteristics of Planting Material in Poly Bags with Coir Pith

A total of 66% per cent sprouting was obtained in nursery bed with rhizome as planting material. Sprouting was initiated in just 24 days and extended up to 42 days for completion. Maximum percent of sprouting was observed on 24 day with a record of 14 sprouting.

62% sprouting was obtained in nursery bed when aerial shoot cuttings were used as planting material. Sprouting was initiated in just 21 days and was completed in 34 days. Maximum percent of sprouting was observed on 24th day with a record of 19.23% sprouting.

5.1.3. Sprouting Characteristics of Planting Material in Poly Bags with Vermi compost

A total of 14 per cent sprouting was obtained in nursery bed with rhizome as planting material. Sprouting was initiated in just 24 days and extended up to 30 days for completion. Maximum percent of sprouting was observed on 28th day with a record of 8% sprouting.

40% sprouting was obtained in nursery bed when aerial shoot cuttings were used as planting material. Sprouting was initiated in just 23 days and was completed in 36 days. Maximum percent of sprouting was observed on 26th day with a record of 13.46% sprouting.

5.1.4. Sprouting Characteristics of Planting Material in Poly Bags with Sterilized Potting Media

A total of 31 per cent sprouting was obtained in nursery bed with rhizome as planting material. Sprouting was initiated in just 18 days and extended up to 51 days for completion. Maximum percent of sprouting was observed on 43rd day with a record of 8% sprouting.

38% sprouting was obtained in nursery bed when aerial shoot cuttings were used as planting material. Sprouting was initiated in just 24 days and was completed in 36 days. Maximum percent of sprouting was observed on 30th day with a record of 4% sprouting.

5.1.5. Sprouting Characteristics of Planting Material in Poly Bags with Potting Mixture

A total of 66% per cent sprouting was obtained in nursery bed with rhizome as planting material. Sprouting was initiated in just 16 days and extended up to 30 days for completion. Maximum percent of sprouting was observed on 24th day with a record of 26% sprouting.

24% sprouting was obtained in nursery bed when aerial shoot cuttings were used as planting material. Sprouting was initiated in just 24 days and was completed in 31 days. Maximum percent of sprouting was observed on 24th day with a record of 12% sprouting.

5.1.6. Sprouting Characteristics of Aerial Soot Cuttings in Root – Trainers

Vegetative aerial shoot cuttings were made into suitable size and have been planted in the prefilled root trainers with coir pith compost and soil. Buds arise within 15 – 20 days, which require frequent watering for maintaining the propagules fresh (Fig. 5.3.a & Fig. 5.3.b). The root trainers were maintained in specially devised glass chambers inside which an average temperature ranges 25°C – 35°C and an average humidity that ranges 70% - 85% throughout the study period. The rooted stem cuttings reach an average height of 20cm in 30 days by the time they are ready for out planting. Leaves can also be harvested at this stage as raw drug for medicinal preparations.

Fig. 5.3.a. Stem cuttings in root trainers   Fig.5.3.b. Rooted cuttings with sprouts

Vegetative aerial shoot cuttings were made into suitable size and have been planted in the prefilled root trainers with coir pith compost and soil. Buds arise within 15 – 20 days, which require frequent watering for maintaining the propagules fresh. The root trainers were maintained in specially devised glass chambers inside which an average temperature ranges 25°C – 35°C and an average humidity that ranges 70% - 85% throughout the study period.

5.2. Growth Performance of Propagules in Various Potting Media

5.2.1. Growth Performance of Propagules in raised soil in Nursery

Growth data were gathered at an interval of 15 days. For standardization purpose data were gathered from 25 plants in each sample media. The average height in length, number of leaves, number of nodes and leaf length are given in table 1. Within first 45 days of growth plants attain an average height of 10.5 cm with 7-15 leaves and 6-15 nodes (Table 5.1).

Table 5.1. Growth data of Propagules in raised soil in Nursery

Period of Observation	Average height (cm)	Average No of leaves	Average No of nodes	Average leaf width (cm)	Average leaf length (cm)
Period 1	5.18	4	6	3.3	3.45
Period 2	5.26	4	9	3.43	7.55
Period 3	10.5	12	13	3.4	8.5

5.2.2. Growth performance of Propagules in Poly Bags with Coir Pith

Within first 30 days of growth plants attain an average height of 4.5 cm with 2-8 leaves and 3 - 4 nodes. The data generated are given in table 5.2.

Table 5.2. Growth data of Propagules in Poly Bags with Coir Pith

Period of Observation	Average height (cm)	Average No of leaves	Average No of nodes	Average leaf width (cm)	Average leaf length (cm)
Period 1	3.41	3	4	2.72	5.64
Period 2	4.5	4.75	6.17	2.9	5.81

5.2.3. Growth performance of Propagules in Poly Bags with Vermi Compost

Within first 30 days of growth plants attain an average height of 6.4 cm however the maximum height observed was 10.5 cm. with 2-10 leaves and 4 -10 nodes. The data generated are given in table 5.3.

Table 5.3. Growth data of Propagules in Poly Bags with Vermi Compost

Period of Observation	Average height (cm)	Average No of leaves	Average No of nodes	Average leaf width (cm)	Average leaf length (cm)
Period 1	3.218	3	5	2.93	4.93
Period 2	6.4	6	8	3.73	8.3

5.2.4. Growth performance of Propagules in with Sterilized Potting Media

In 30 days of growth, plants attain an average height of 9.73 cm with 4 - 8 leaves and 4 - 10 nodes. The data generated are given in table 5.4.

Table 5.4. Growth data of Propagules in Poly Bags with Sterilized Potting Media

Period of Observation	Average height (cm)	Average No of leaves	Average No of nodes	Average leaf width (cm)	Average leaf length (cm)
Period 1	3.97	3	5	2.76	5.32
Period 2	9.73	6	7	3.58	7.4

5.2.5. Growth performance of Propagules in with Sterilized Potting Media

In 30 days of growth, plants attain an average height of 3.84 cm with 2 - 6 leaves and 3 - 7 nodes. The average data on different parameters considered are given in table 5.5.

Table 5.5. Growth data of Propagules in Poly Bags with Potting Mixture

Period of Observation	Average height (cm)	Average No of leaves	Average No of nodes	Average leaf width (cm)	Average leaf length (cm)
Period 1	2.8	2	4	1.8	4.6
Period 2	3.84	3	4	2.95	5.22

5.2.6. Growth performance of Propagules in Root- trainers

In 30 days of growth, plants attain an average height of 7.1 cm with 3 - 5 leaves and 3 - 7 nodes. The average data on different parameters considered are given in table 5.6.

Table 5.6. Growth data of Propagules in Root - trainers

Period of Observation	Average height (cm)	Average No of leaves	Average No of nodes	Average leaf width (cm)	Average leaf length (cm)
Period 1	4.2	3	5	3.01	4.52
Period 2	7.1	5	7	3.83	7.4

5.3. Biomass study on propagules grown in Different Growing Media

In order to generate data on biomass content in propagules from different potting media, one plant each from each sample group was selected and its wet as well as dry weights were gathered using electronic balance.

Plant materials were dried using hot air oven in 65 – 70°C for 72 hours. All the data generated are given in table 5.7.

Table 5.7. Wet weights and Dry weights of samples grown in different growing media

Sl. No:	Plant code	WET WEIGHTS (g)					DRY WEIGHTS (g)				
		whole plant weight (g)	Shoot weight (g)	Rhizome weight (g)	Additional Shoot weight (g)	Root weight (g)	Root weight (g)	Rhizome weight (g)	Shoot weight (g)	Additional Shoot weight (g)	whole plant weight (g)
1	B1A28	50.82	11.85	33.42	1.07	3.87	1.0399	1.9354	0.8627	0.0516	3.8896
2	B2A46	8.17	6.47	0.61	0.45	0.45	0.1067	0.0381	0.4999	0.0263	0.671
3	B3A5	38.77	30.5	8.26		0.41	0.0507	0.2434	1.3238		1.6179
4	B4A7C	3.69	3.18	0.20	0.13	0.14	0.0201	0.0133	0.2172	0.008	0.2586
5	B4A2R	36	10.05	25.18		0.75	0.1155	1.5084	0.5933		2.2172
6	B4A2P	47.33	14.62	24.04	7.87	0.84	0.0675	0.8487	1.889	0.2663	2.0715
7	$B_6$5CC	5.89	4.01	0.53		1.24	0.0721	0.0535	0.4641		0.5897
8	$B_6$8CV	38.61	20.37	2.01	12.01	3.86	0.2671	0.1006	1.099	0.2206	1.6873
9	$B_6$6RC	56.25	14.37	35.78	0.12	5.76	0.6126	2.3573	1.2799	0.0066	4.2564
10	$B_6$3CM	31.05	9.42	2.52	10.21	8.36	0.5047	0.1826	0.3875	0.6538	1.7286
11	$B_6$4RV	38.8	11.91	23.7	0.16	2.88	0.1400	1.3315	0.6342	0.009	2.1147
12	$B_6$22RM	41.45	10.05	28.05	0.37	2.83	0.1195	1.2323	0.5193	0.0130	1.8841
13	$B_6$2RCnt	43.79	10.56	29.60	0.02	3.51	0.2705	1.4550	0.5937	0.0065	2.3257
14	$B_6$17CCn	13.49	6.86	0.47	0.04	1.70	0.1236	0.0368	0.3133	0.0017	0.4754

In order to generate data on biomass content in propagules from different potting media, one plant each from each sample group was selected and its wet as well as dry weights were gathered using electronic balance. Plant materials were dried using hot air oven in 65 – 70°C for 72 hours. All the data generated are given in table 5.7.

5.4. Conclusion

From the data obtained it can be inferred that *Costus pictus* plant can be cultivated easily in tropical climatic conditions. Aerial shoot cuttings and rhizome may be the most suitable planting material as there is no much variation in sprouting percentage was observed. Potting media such as Coir pith and vermi compost are suitable for growing purpose other than the standard potting mixture.

References

Baker, J. G., 1890 – 1892. Scitaminae in J. D. Hooker. *Flora of British India*, Vol.6: 198-264. London.

Bailey, L. H. 1920 (revised).*The nursery manual*. Macmillan, New York.

Baltet, C. 1910.*The art of grafting and budding*. 6th ed. London: Crosby Lockwood. (quoted by Hottes, 1922).

Davidson, H., R. Mechlenburg, and C. Peterson. 2000. *Nursery management*. 4[th]ed. N.J, Prentice Hall, Upper Saddle River.

Fisher, C.F.C., 1928. Zingiberaceae. In J.S. Gamble. *Flora of the presidency of Madras*, Pt. 8: 1478-1493. London.

Fuller, A.S. 1887. *Propagation of Plants* (quoted by Hottes, 1922).

Gamble, J.S., 1916 – 1935. *Flora of the presidency of Madras*. Pts. 1-11. London.

Harlan, J.R. 1992. *Crops and man*. 2nd ed. Madison, Wia. Amer. Soc. Of Agron., Inc. Crop Science of America. America

Hartmann, H. T., A.M. Kofranef, V.E. Rubatsky, and W.J. Flocker. 1988. *Plant Science: Growth, development and utilization of cultivated plants*. 2nd ed. N.J. Prentice Hall. Englewood Cliffs.

Janick, J., W. Shery, F.W. Woods, and V.W. Ruttan. 1969. *Plant Science*. W. H. Freeman, San Francisco.

Kerala Land Use Board, 1997. Govt. Press. Trivandrum

Kress, W.J., Linda M. Prince & K.J. Williams, 2002. The Phylogeny and new classification of the gingers (Zingiberaceae): Evidence from molecular data. *Amer. J. Bot.* 89 (11): 1682-1696.

Larson, K., J.M. Lock, H. Mass and P.J.M. Mass. 1998. Zingiberaceae: In:Kubitzki (ed.), *The families and genera of vascular plants*. 4. Springer verlag, Berlin Pl. 4774-495.

Mass, P.J.M., 1972. Notes on Asiatic and Australian Costoideae (Zingiberaceae) *Blumea* 25: 543-549.

Nakai, T. 1941.*Notulae ad Plantas Asiae Orientalis*. XVI. J. Jap. Bot. 17: 189-210.

Rama Rao, M. 1914.*Flowering Plants of Travancore*. Govt Press. Trivandrum.

Reed, H.S. 1942. *A short history of the plant sciences*. The roland Press Co. New York.

Rheede tot Drakenstein, H. A. van. 1678-1693. *Hortus Indicus Malabaricus*. Vols. 1 – 12. Amsterdam.

Sabu, M. 2006.*Zingiberaceae and Costaceae of South India*. Calicut, India.

Sasidharan, N. 2007.*Flowering plants of Kerala*. CD No.6. Kerala Forest Research Institute, Peechi.

Sasidharan, N. 2004 *Biodiversity documentation for Kerala* Part 6: Flowering plants. Kerala Forest Research Institute, Peechi.

Shyam, A.T.S., 2007. *Ground water booklet of Ernakulam District Kerala State.*Central Ground Water Board, Kerala region, Kedarom, Trivandrum.

Solbrig, O.T., and D.J. Solbrig.1994. *So shall you reap.Farming and crops in human affairs*. Island Press. Washington, D.C.

Ward, N.B. 1842. *On the growing of plants in closely glazed cases*. 2nd ed. J. van Voorst, London.

Printed by Books on Demand GmbH, Norderstedt / Germany